Bibliographic information published by the German National Library:

The German National Library lists this publication in the National Bibliography; detailed bibliographic data are available on the Internet at http://dnb.dnb.de .

Imprint:

Copyright © 2018 GRIN Verlag
Print and binding: Books on Demand GmbH, Norderstedt Germany
ISBN: 9783668667587

This book at GRIN:

https://www.grin.com/document/416773

William Fidler

Structure and Expansion of a Universe whose Vacuum is modelled as a Set of identical, bi-modal, frequency-quantised, simple quantum harmonic oscillators.

GRIN Verlag

The structure and expansion of a universe obeying Hubble's law and whose vacuum is modelled as a set of identical, bi-modal, frequency-quantised, simple quantum harmonic oscillators.

W M Fidler

Abstract

The 'fabric' of the model presented here consists of identical, contiguous cubes of space, each of which is occupied by a simple quantum harmonic oscillator. A possible structure is proposed for the ensemble of cubes, and it is shown also that other configurations, on all scales can exist in the ensemble, even although the members of that collection are absolutely identical.

The expansion of space is considered initially from the standpoint of a universe which is not accelerating, and equations are developed which relate to this 'coasting' condition, and, in which the Hubble parameter, H, is shown to be a function of time. These equations are extended by the inclusion of the deceleration parameter of cosmology, q, treating q as a parameter.

The observations of Schmidt, Riess et al* and Perlmutter et al** have shown that approximately five billion years ago the universe experienced a jerk and went from a decelerating state to one of acceleration. From these findings it is argued that the deceleration parameter is not constant, but is a function of time. A distribution for q of the form: $q = e^{k(t_0 - t)} + B$ is proposed, where k and B are constants and the subscript 0 on t refers to time 'now'. This guess proves to be reassuringly productive for, with the appropriate boundary conditions, it captures the deceleration phase before the jerk and the coasting condition through which the universe must pass in order to attain the accelerating state. In addition, the equation shows that the acceleration is increasing after the jerk. This is consonant with the observations of the above-mentioned experimenters.

The magnitude for q favoured by some theorists is ½. From the first extended equations (see the text for explanation) it is then shown that this q is located in time $^2/_{3H_0}$ before the present, where H_0 is the current magnitude of Hubble's parameter $(2.38 \ x \ 10^{-18} \ s^{-1})$. However, it is seen that this time is determined from the set of extended equations mentioned above, where q is considered to be a parameter.

By incorporating the expression proposed for q to form the second extended equations it is shown that q = ½ is located at a time in the past close to $^{0.925}/_{H_0}$. Two methods are outlined which will permit the speed of expansion of space at the jerk to be determined, together with a rough estimate of that speed.

It is shown that the Hubble parameter, which should, strictly, be termed 'Hubble's function' in view of its time dependence can be interpreted in two different ways, one of which is that Hubble's function is equal to one third of the volumetric strain rate of space.

It is concluded, for this model that the universe is infinite and observation of such is observer-centred. Further, it is argued that each observer can, in principle see a different part of the universe, but, due to the current size calculated for the part of the universe examined here, spatially, but closely- separated observers do not possess sufficient precision to distinguish any differences in their respective far-field observations.

List of Contents

Introduction

It is accepted generally that the shape and size of the universe is unknown, or, indeed if it is even finite. Calculation by others of the size of the universe which, in principle, is capable of being observed yields a spherical shape with a diameter in the region of 93×10^9 light years. It is of interest to note that, for a universe containing 10^{80} baryons and modelled as a collection of contiguous cubes, it is calculated in **((r2))** that the side of this array is 76×10^9 light years. The volume of this cube may be contained within a sphere of diameter, 94.3×10^9 light years. In addition, the boundaries of current models of large-scale galaxy formation, simulated numerically in supercomputers, are the sides of very large cubes.

The speed of the expansion of space has only been confirmed at great distances from the point of observation and was determined experimentally by Edwin Hubble. Hubble established a linear relationship between the speed of the expansion at a point in space and the distance of an observer from that point. The parameter which reduces the proportionality between the speed of expansion and the distance is called, probably incorrectly, Hubble's constant, for it is thought to be a function of time; further, the magnitude of the parameter has been the subject of much debate over many decades. The latest estimate seen by the author for Hubble's parameter is $\frac{73.8 \times 10^3}{3.28 \times 10^6}$, neglecting the error band; the units of this quantity are km/s/light year. When used in this form the distance to the point of observation should be expressed in light years. The magnitude of the parameter, if the distance is expressed in metres, is, $2.38 \times 10^{-18} \, s^{-1}$.

This hybrid model, which employs results from relativity and quantum mechanics, is based entirely upon average conditions, takes no account of the gravitational effect (if any) of the mass/energy which may occupy space. Indeed, at the current time, from the data garnered by the Planck mission, it has been determined that space contains, on average, slightly less than the equivalent of 6 protons per cubic metre, composed of baryonic matter, Dark matter and what is sometimes called the Dark energy, but in **(r1)** is shown to be related to the ground state of a quantum harmonic oscillator. It is acknowledged that the first two of these have effects on space opposite to that of the third.

It may be argued that, since the current mass/energy density is so low, it is not necessary to take account of any countervailing gravitational mechanism at the present time, and hence the expansion of space at all of the levels in this model may be considered.

Indeed, since this model and also that developed in **(r1,r2,r3)** considers only average conditions, then the highly diffuse conditions which are concomitant with the averaging process are extant on all scales of these models.

A stacking structure for the universe

The model of the universe presented here consists of a set of identical, simple, frequency-quantised, quantum harmonic oscillators, each of which occupies a cubic space of side equal to the prevailing wavelength; we describe the cubic space as an enclosure and designate such an enclosure as a fundamental cube. Aspects of the model used here are discussed at length in **(r1)**, **(r2)** and **(r3)**.

Now, a simple harmonic oscillator oscillates only in one direction, and a simple quantum harmonic oscillator is no exception. If the cubes are stacked together at random to form a larger cubical structure then it is possible that sub-structures may be formed where, amongst other things, there are predominant directions of oscillation due to cubes of the same oscillatory direction meeting along a face, the normal to which is in the direction of the oscillation. This would imply a preferred direction in such instances, and would constitute a wave train of length proportional to the number of cubes.

If it is posited that space, on large scales, is very closely isotropic, then configurations such as this violate isotropy. It follows that if space is to be closely isotropic, then we must consider how structures such as the above may be minimised, or even avoided, and also possible arrangements of cubes which will render space isotropic. An overall structure in which space is both homogeneous and isotropic is characteristic of a Friedman-Lemaitre-Robertson-Walker (FLRW) universe.

Within the constraints of the foregoing we now proceed to investigate cubical structures. To facilitate this we assign colours, red (R), yellow (Y) and blue (B) to the three possible directions of oscillation with respect to a set of Cartesian axes located at the vertex of a cube. It should be noted that these distinctions only find meaning in the context of their assembly into larger cubical structures. It is helpful to envisage the now distinct cubes as a set of child's coloured building blocks.

On the accompanying diagram sheet the fundamental cubes and all combinations thereof are, for clarity, shown in plan view.

The basic elements are as shown in Fig1.

Figs1(a),(b),(c) and (d) show a set of arrangements which all will form an '8' cube. It is immediately evident that all of the arrangements shown conform to the requirement that no two cubes of the same colour meet along a face.

However, we see, for example, in (a) that, whilst cubes of all of the colours are represented, there are twice as many cubes of one colour as there is of cubes of the other colours, whilst for the arrangements shown in Figs (b),(c) and (d), in each example, one of the colours is not present.

R
Y
B

Fig 1

Y R
B Y
+
B Y
Y R
(a)

Y R
R Y
+
R Y
Y R
(b)

B R
R B
+
R B
B R
(c)

Y B
B Y
+
B Y
Y B
(d)

B Y B
Y R Y
B Y B
(a)
+
Y R Y
R B R
Y R Y
(b)
+
R B R
B Y B
R B R
(c)

Fig 2

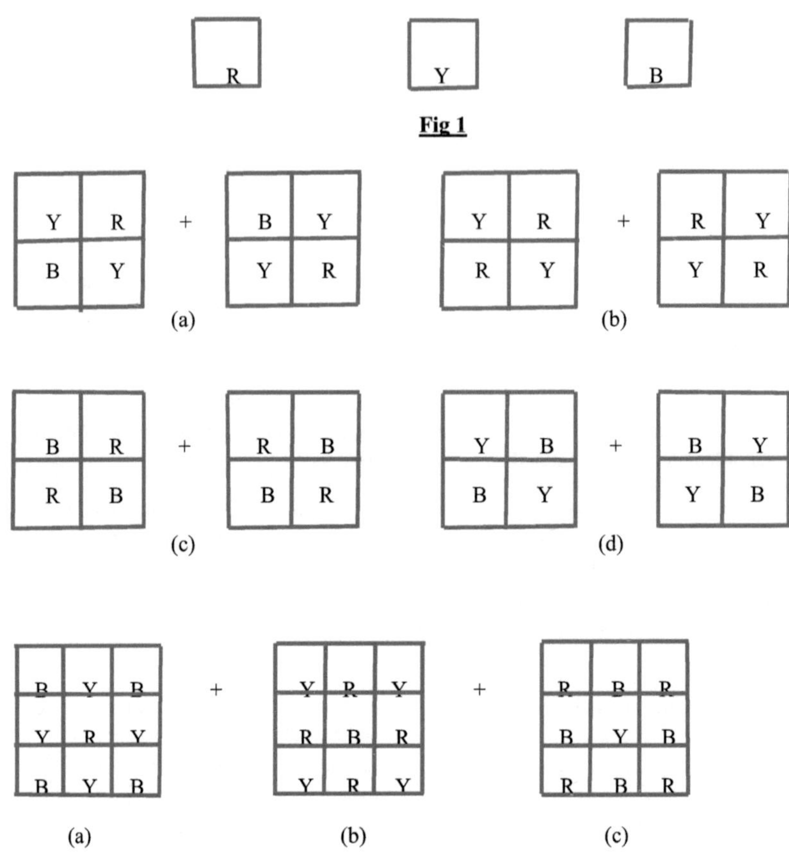

Whilst the arrangements of cubes shown accord with the stipulation that no cubes of the same colour may meet along a face, the resulting structures are considered to be non-isotropic for the reasons given above.

Consider now the arrangements shown in Fig2. It is obvious that when the individual planes of cubes are stacked, a '27' cube will result which accords with a now stronger requirement that, for any cubical structure formed from the fundamental cubes, no cubes of the same colour may meet along a face, irrespective of how the planes of cubes are stacked, but also, that the structure contains the same number of cubes of each colour. For brevity, we describe such a structure, and all other structures which display these characteristics, as 'isotropic cubes'. The system examined here displays characteristics typical of macroscopic structures which are comprised of microscopic elements, in that, on a small scale there may be significant differences between structures, whilst, on large, or very large scales, differences may not be discernible. It may be inferred that the entropy increases in going from small, to large, structures. This is analogous to the situation in a macroscopic collection of gas molecules.

As a general rule, it follows that one of the conditions for isotropy of any composite cube is that the cube contain an equal number of the different colours, and hence, since the number of fundamental cubes in any cubical structure is given by the cube of the number of cubes in the length of a side, then to accord with one of the above requirements, the number of cubes in the length of the side of a cube must be exactly divisible by three---otherwise the structure cannot even begin to be considered to be isotropic. If we write out some of the set of natural numbers, then, each number may be taken to represent the number of fundamental cubes in the length of the side of a larger cube. Also, as shown below, the numbers shown in parentheses correspond to the cube root of the number of fundamental cubes which, we assume, may be assembled into an isotropic cube. Of course, to accord with the requirements for an isotropic cube the structure so formed must also conform to the requirement that no cubes of the same colour may meet along a face.

1,2,(3),4,5,(6),7,8,(9),10,11,(12),,,,,,,,,,,etc.

Further, it is noted that if we form the sums of two consecutive numbers from the set of natural numbers, then, if the sum is exactly divisible by three, none of the numbers of which the sum is composed can be considered to be proportional to the length of the side of an isotropic cube, whereas, if the sum is not exactly divisible by three then one of the numbers is proportional to the length of the side of an isotropic cube.

We note that all of the foregoing constitutes an explanation of how, despite the averaging nature of the model, non-uniformities may arise in an otherwise, featureless, and very large array of fundamental cubes. We make no attempt to describe, in detail, these irregularities, other than to speculate that their distribution might be Gaussian. Further consideration of all of the above is left to those whose expertise lies in the field of combinatorics.

The expansion of space

The model developed in **(r1)**, **(r2)** and **(r3)** consists essentially of a very large set of cubes of space which are all of the same size at any given time, t. The length of the side of a cube is the wavelength determined from the formula, $v\,l = c,$ where v is the frequency of oscillation of the set of quantum harmonic oscillators of which the vacuum is predicated to be comprised. The cubes are stacked in the manner described earlier. When we speak of changes occurring over some time interval, we are, in effect, examining differences between two snapshots (frames---hyperframes?) of the ensemble of cubes. A massless test point embedded in space will be observed to have changed its position when the two frames are compared, not because the point has moved through space, but because space has moved through time. The perceived speed of the test point is, in fact, the speed of expansion of space. We cannot directly observe the expansion of space but must observe the motion of co-moving objects embedded within it.

Consider such an object embedded in a plane situated at a distance, R, from the origin, O, where distance is measured along the normal to the plane, as shown in Fig3. At time, t, (frame1) let the object be moving away from O at speed, v. In a time interval, dt, (frame2) the object will be perceived to have moved a distance, dR, and in this new position the speed will seem have changed to, $v \pm dv$. Hence, the mean velocity during this interval is, $v \pm \dfrac{dv}{2}$.

Now, $dR = \left(v \pm \dfrac{dv}{2}\right)dt = v.\,dt$ to first order, and hence, $\dfrac{dR}{dt} = v.$ The distance, R, is now regarded as one half of the side, L, of a composite cube centred on O.

The volume of the cube is: $L^3 = N_{enc}.\,l^3$, where N_{enc} and l are the number of fundamental cubes in the composite cube and l is the length of the side of a fundamental cube, respectively; the subscript 'enc' derives from the earlier statement that a fundamental cube is described as an enclosure.

Hence, $L = \sqrt[3]{N_{enc}}\,l$, and, differentiating wrt time, we get:

$$\frac{dL}{dt} = \sqrt[3]{N_{enc}}\,\frac{dl}{dt} \text{ ----------- (1).}$$

Differentiating $v\,l = c,$ wrt time; $\dfrac{dl}{dt} + \dfrac{l}{v}\dfrac{dv}{dt} = 0$. We may then write equation (1) as :

$$\frac{dL}{dt} = -\sqrt[3]{N_{enc}}\,\frac{l}{v}\,\frac{dv}{dt} \text{ ------------ (2).}$$

Now, $L = 2R = \sqrt[3]{N_{enc}}\,l$, and $\dfrac{dL}{dt} = 2\dfrac{dR}{dt} = 2v.$ Hence, we may write equation (2)

as: $$v = -R.\frac{1}{v}\,\frac{dv}{dt} \text{ ------------------ (3).}$$

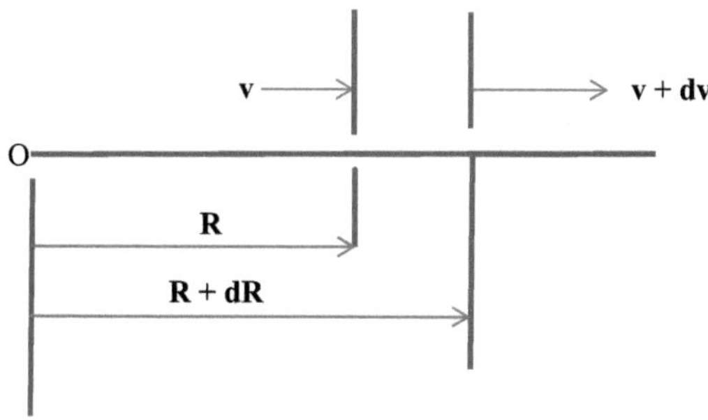

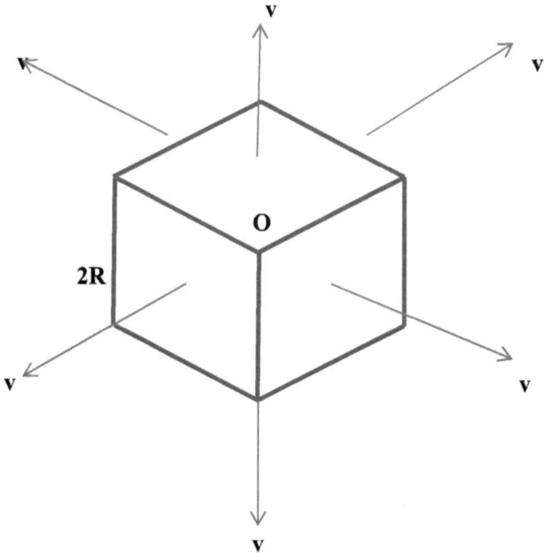

Fig 3

The model of the vacuum developed here is such that, at any time, t, v and dv/dt have the same values for all fundamental cubes, hence, v/R has the same value at the time, t, for all structures formed from these cubes, irrespective of size and orientation. Moreover, equation (3) is an expression for the rate of expansion of space in three mutually perpendicular directions, for, the orientation of the plane is not specified, other than that it is one side of a cube, and v is positive.

It is important to stress that, for this model, whilst v/R is constant throughout all of space at any given time, this must not be taken to imply that the ratio is constant for all time.

Further, equation (1) may be written: $dR/dt = \sqrt[3]{N_{enc}}\; dr/dt$, where $r = l/2$.

We may also write this equation as: $v/\breve{v} = \sqrt[3]{N_{enc}}$, where $\breve{v}$ is the speed of expansion of the space enclosed by a fundamental cube; it is seen that this equation is quantised, for, $\sqrt[3]{N_{enc}}$ may only take values from amongst the set of natural numbers.

A relationship is now sought between v and R.

Such a relationship exists in the form of the Hubble formula, $v = R.H$, where H is the Hubble parameter. At any given time, the formula has been validated experimentally only at great distances from the Earth, and upon observations of the motion of gravitationally-isolated structures. Moreover, the Hubble parameter may be a function of time. For this model, great distances are associated with very large cubes, but, as argued above, for this model the formula is applicable to a fundamental cube or any cubical structures composed thereof. Hence, at the current time we may apply the Hubble formula using the current value of the Hubble parameter at any of the scales of the model.

We may then write equation (3) as: $\dfrac{1}{v}\, dv/dt = -H$ ----------------- (4) .

The Hubble parameter may also be written $H = \dfrac{1}{a}\, da/dt$, where a is the scale factor.

Hence, we may write $\dfrac{1}{v}\, dv/dt = -\dfrac{1}{a}\, da/dt$. Now, $\dfrac{1}{v}\, dv/dt = -\dfrac{1}{l}\, dl/dt$. From all of this we obtain $\dfrac{1}{a}\, da/dt = \dfrac{1}{l}\, dl/dt$, which, upon integration shows that the scale factor, a, and the wavelength, l, are related by the expression: $a/l = a\ constant$.

It then follows that: $a/a_o = l/l_o$ ------------------- (5), where the subscript 'o' denotes some reference condition.

In (r3) the temporal frequency distribution was assumed to be linear throughout the second free vacuon phase. It may be seen from a cursory examination of equation (4) that this is cannot be correct. We obtain a more realistic distribution by integrating equation (4). However, this will do little to advance the analysis, unless a relationship between the Hubble parameter and time can be determined.

Integrating (4) $\int_{v_o}^{v} dv/v = -\int_{t_o}^{t} H\, dt$, giving: $v/v_o = \exp\left[-\int_{t_o}^{t} H\, dt\right]$ -------- (6).

Equation (6) shows, for this model, that, ostensibly, the frequency ratio is an unknown exponential function of time. In light of this, the assumption, in (r3) of a linear variation throughout the second free vacuon phase is not correct and was only adopted at the time for lack of further analysis, such as that performed above. Indeed, we see that this must be the case if we rewrite equation (4) as $dv/dt = -v H$ -------------------- (7).

.If it is assumed that the frequency distribution from now to the inception of the universe is given by equation (6), integrated as if the Hubble parameter is constant and equal to the magnitude stated earlier, and, taking $v = 0.515 \times 10^{28}$ Hz, $v_o = 2.681 \times 10^{12}$ Hz, (see (r2)), there results an age for the universe far older than the currently- accepted value.

It has been noted earlier that, according to the model developed in (r2), the universe has the form of a cube, currently of side 76 billion light years; hence, using the magnitude of the Hubble parameter quoted earlier, the speed, v, of the expansion at the edge of the cube is calculated to be: $\mathbf{38 \times 10^{9} \times 9.454 \times 10^{15} \times 2.38 \times 10^{-18} \times 10^{-3} = 855 \times 10^{3}\ km/s}$, which is **2.85 times the speed of light**. If we express the currently accepted age of the universe (13.725 billion years) as a distance in light years then the speed of expansion at this distance is determined to be 1.03 times the speed of light. It has been reported that some of the galaxies at the furthest limits of observation are receding at speeds close to that of the speed of light.

 It should be remarked that the motion of space at greater than luminal speed does not violate Special Relativity for that theory relates only to motion through space.

Had these speeds been calculated with the magnitude of the rate of change of frequency during the second free vacuon phase based upon that determined in (r3) then the speeds of expansion would be approximately four orders of magnitude less than those shown above, and consequently, in substantial conflict with observation.

From the expression: $dR/dt = \sqrt[3]{N_{enc}}\ dr/dt$ we may determine the speed of expansion of space at the level of a fundamental cube by using the Hubble formula and setting N_{enc} equal to unity; as noted previously, the above formula is applicable at all scales of this model.

The wavelength corresponding to the current frequency of the vacuum has been determined in previous work, viz. 1.118×10^{-4} m; the current speed of expansion of space at the level of a fundamental cube, $\breve{v}$, is then given by the Hubble formula as:

$$\breve{v} = \frac{1.118 \times 10^{-4}}{2}\ 2.38 \times 10^{-18} = 1.33 \times 10^{-22}\ m/s.$$

We have already shown that the rate of expansion of space, v, at any given distance from an observer may be written: $dR/dt = \sqrt[3]{N_{enc}}/2 \; dl/dt$, which, since $dl/dt = -c/v^2 \; dv/dt$,

becomes: $v = -\sqrt[3]{N_{enc}}/2 \; c/v \left(1/v \; dv/dt\right) = \sqrt[3]{N_{enc}}/2 \; cH/v$ by virtue of equation (4).

Differentiating the last expression wrt time, we get:

$$dv/dt = \sqrt[3]{N_{enc}}/2 \; c \; d/dt \left(H/v\right) \text{------------------- (8).}$$

Hence, $dv/dt = \sqrt[3]{N_{enc}}/2 \; c \left(1/v \; dH/dt - H/v \left(1/v \; dv/dt\right)\right)$, which, using equation (3) reduces to :

$$dv/dt = \sqrt[3]{N_{enc}}/2 \; c/v \left(dH/dt + H^2\right) \text{---------------- (9).}$$

A dimensionless measure of the deceleration parameter, q, of the cosmic acceleration of the expansion of space in a FLRW universe is defined by: $q = -a\ddot{a}/\dot{a}^2$, and, using $H = \dot{a}/a$ it is easy to show that q may be written in terms of Hubble's parameter, viz.

$$dH/dt + H^2 = -qH^2.$$

Now, equation (9) may be set to equal to zero only by setting the bracketed term therein equal to zero. This is equivalent to setting the deceleration parameter equal to zero.

Consider the vanishing of the bracketed term, i.e. $dH/dt + H^2 = 0$ -------- (10).

We have a simple, first order differential equation which relates the variation of the Hubble parameter with time, and which, in this form, is applicable only to a non-accelerating (i.e. coasting) universe.

Separating the variables and integrating: $\int_{H_o}^{H} dH/H^2 = \int_{t}^{t_o} dt$.

$\therefore \left[1/H\right]_{H}^{H_o} = (t_o - t)$, from which we obtain the expression, for this model, for the temporal distribution of the Hubble parameter for a coasting universe, viz.

$$H = H_o/[1 - H_o(t_o - t)] \text{-------------- (11).}$$

The subscript ' o' denotes the present time.

In the future, the time difference, $t_o - t$ is negative and from (11) it is then seen that, in the future the Hubble parameter will be smaller than its current magnitude, whereas, in the past, the time difference will be positive and the Hubble parameter will be greater than is currently the case.

It may also be inferred that, in the future, at least for a coasting universe, an observer looking out to a given distance will observe test points carried with the so-called 'Hubble Flow' to be moving more slowly than current observations of speed at this distance.

Examining equation (11) we see, at some time in the past that the denominator becomes zero (and H, thus, infinite). This time, measured from the present is given by the inverse of the current magnitude of the Hubble parameter ($2.38 \ x \ 10^{-18} \ s^{-1}$). Hence, the time interval is calculated to be 13.332 billion years, and is the well-known, Hubble time. This is close to the currently-accepted age of the universe. Indeed, the current magnitude of the Hubble parameter, calculated from the inverse of the currently-accepted age of the universe is $2.312 \ x \ 10^{-18} \ s^{-1} \equiv 71.7 \ ^{km}/_{s \ Mpc}$, and this is well within the error band established experimentally for this quantity.

From equation (11) it is easy to show that the time interval, $t_o - t$ for a coasting universe, is equal to the difference between the respective Hubble times.

If the temporal difference is greater than 13.725 billion years, with the reference Hubble parameter set to its currently-accepted value, then the Hubble parameter becomes negative.

We are now in a position to determine the temporal frequency distribution for a universe where the variation of the Hubble parameter is given by equation (11)

Equation (6) may now be written $^{v}/_{v_o} = \exp\left[- \int_{t_o}^{t} {^{H_o}}/_{[1 \ - \ H_o(t_o \ - \ t)]} \ dt\right].$

Performing the integration gives: $^{v}/_{v_o} = {^{1}}/_{[1 \ - \ H_o \ (t_0 \ - \ t)]}$ ------------ (12).

From this we note that, despite what was claimed earlier, in this instance, the frequency ratio is not exponential, but is hyperbolic in form.

It is seen that in the future, the time difference $(t_o - t)$ is negative and hence the frequency reduces with time, whereas, in the past, $(t_o - t)$ is positive and the frequency rises 'into the past'.

Using equations (5), (11) and $v \ l = c$ we may write an extended form of (12), viz.

$$^{v}/_{v_o} = {^{H}}/_{H_o} = {^{a_o}}/_{a} = {^{l_o}}/_{l} = {^{1}}/_{[1 \ - \ H_o \ (t_0 \ - \ t)]} \ \text{------------ (13).}$$

It is noted that equation (6) could have been derived by integrating the differentiated form of Hubble's formula wrt time, treating the Hubble parameter as a function of time, and setting the LHS set equal to zero. However, this action, in isolation would not have yielded the rich harvest shown in equations (13).

We now consider a consequence of the assumption that the linear acceleration of space is zero.

As has been shown previously, the speed of expansion of space, v, is given by:

$$v = H\,R; \text{ Now, } R = \sqrt[3]{N_{enc}}/_2\; l, \text{ hence, } v = H\,\sqrt[3]{N_{enc}}/_2\; l.$$

From equations (13) it is seen that $H\,l = H_0\,l_0$

It then follows that, $v = \sqrt[3]{N_{enc}}/_2\; H_o l_o$ ---------------- (13a).

Hence, the speed of expansion of the space of a coasting universe at any time, t, is, as expected, independent of time. For this model we see that there is no need to determine a value for the Hubble parameter at a time other than 'now' in order to calculate the speed of expansion of space at some time other than 'now', for a given value of N_{enc} .

We have already calculated the current speed of expansion at the edge of the universe to be 2.85 times the speed of light. If the universe is coasting at the present time and also at all past times (and this has been determined by observation not to be the case) then, according to equation (13a), this should also be the speed of expansion of the spatial boundary of the universe at inception, when, as calculated in **(r2)**, the oscillators comprising the universe occupied a cube of side 4×10^{-5} light years, and the frequency of oscillation of the vacuum was determined to be $0.515 \times 10^{28}\ Hz$, both of these being approximate and predicated upon there being 10^{80} baryons in the universe at the current time.

Denoting variables at the inception of the model of the universe developed in **(r2)** by the subscript, 'i', then, from equation (13) we have:

$$H_i = H_o\; {}^{v_i}/_{v_o} = 2.38 \times 10^{-18}\,.\,0.515 \times 10^{28}/_{2.681 \times 10^{12}}$$

$$= 4.612 \times 10^{-3}\ s^{-1}.$$

$$\text{Hence, } v_i = H_i.R_i = (4.612 \times 10^{-3} x\, 2 \times 10^{-5} x\, 9.454 \times 10^{15})/_{2.997925 \times 10^8}$$

The result of this calculation shows that v_i = **2.9 times the speed of light**.

Despite the approximate nature of the values employed in the calculation, the result verifies (13a).

It will be noticed by now that the analysis uses differential and integral quantities rather than finite differences and sums, despite the title of the work saying that the frequency of the quantum harmonic oscillator is quantised. The justification for this is that in **(r3)** we set, quite arbitrarily, the magnitude, Δ, of the degree of frequency quantisation, to be that associated with a rise in temperature of approximately 1K at the inception frequency of the universe described in that work. The frequency quantisation was then calculated to be 1.219×10^{-40} Hz. Given the frequency range covered by the analysis, this magnitude, on any objective basis, may be considered to be a very good approximation to a differential quantity.

It is noted that the gross characteristics of a universe described by equations (13) bear all of the hallmarks of a typical Big Bang model, for, when the time interval is positive and is that of the age of the universe, all of the non-referenced quantities become either zero or infinite in magnitude. We reiterate that no account is taken of the effect of matter/energy density upon the geometry of space at any stage of the analysis. It is seen from equation (13), that to proceed to times in the past, such that the time interval is greater than the inverse of the reference Hubble parameter, results in the relevant quantities becoming negative, and from this it is concluded that time began at the inception of the universe. Indeed, from the last term on the LHS of (13) we see that, under similar circumstances, the wavelength of an enclosure becomes less than zero, and hence it follows that space also began at the inception of the universe. Thus, we are faced with the imponderable singularity of a Big Bang.

It is not our intention here to even attempt to investigate conditions at the inception of this universe.

It is considered apposite, at this juncture to state that the model developed in **(r1)**, **(r2)** and **(r3)** , where, as here, the vacuum consists of a set of identical, bi-modal, frequency-quantised, simple quantum harmonic oscillators, is not plagued by the infinities and zeros of Big Bang models and may be extended indefinitely into the past. However, the constant temporal frequency gradient during the second free vacuon phase was predicated upon the occurrence of a some sort of effect occurring five billion years ago and at a frequency close to the current frequency of the vacuum, and, as mentioned before, this, together with the assumption of the linear temporal frequency distribution would render the current calculated speed of expansion of space to be in error by a factor of roughly four orders of magnitude.

We see that there is no need to assume a temporal frequency distribution for this model, for that distribution is given explicitly by equation (12).

Consider the following simple analysis:

As before, we may express the volume of a cube as: $V = N_{enc}l^3$.

Differentiating wrt time, we get $dV/dt = 3\,N_{enc}l^2\,dl/dt.$

Now, $dl/dt = -l/v\,dv/dt = l\,H,$ using equation (4).

Therefore $\frac{dV}{dt} = 3\,N_{enc}l^3\,H = 3\,VH$ ----------------- (14),

which may also be written $\frac{1}{V}\frac{dV}{dt} = 3\,H.$ ------------ (14a).

It should be noted that the above have been obtained without using any of equations (13). A physical interpretation of equation (14a) is discussed later in the work.

We now proceed to extend the scope of equations (13), with the proviso that the deceleration parameter noted below is considered <u>not</u> to be a function of time.

As stated earlier the deceleration parameter q of cosmology can be shown to be related to Hubble's parameter and may be written: $\frac{dH}{dt} + (1+q)\,H^2 = 0$ ----- (14b).

Treating q as parameter, this is integrated between the limits, H_o and H; t_o and t, to yield:

$$H = \frac{H_o}{[1 - (1+q)H_o(t_o - t)]} \text{------- (14c).}$$

Quite separately from the deceleration parameter we have shown that, $Hv = -\frac{dv}{dt}.$

Hence we may write: $\int_{v_o}^{v} \frac{dv}{v} = \ln \frac{v}{v_o} = \int_{t}^{t_o} H\,dt.$

Giving $\ln \frac{v}{v_o} = H_o \int_{t}^{t_o} \frac{dt}{[1 - (1+q)H_o(t_o - t)]}.$

Using the substitution, $\varphi = (1+q)H_o(t_o - t)$, this may be integrated to give:

$$\frac{v}{v_o} = \frac{1}{([1 - (1+q)H_o(t_o - t)])^{\frac{1}{1+q}}}.$$

The extended form of equations (13), termed the first extended equations, is then:

$$\frac{v}{v_o} = \frac{l_o}{l} = \frac{a_o}{a} = \left(\frac{H}{H_o}\right)^{\frac{1}{1+q}} = \frac{1}{([1 - (1+q)H_o(t_o - t)])^{\frac{1}{1+q}}} \text{------------ (15).}$$

We emphasise that the above set of equations is only valid for $q \neq q(t)$. Further, equations (14c) represent a family of hyperbolic curves with q as parameter, all emanating from $t = t_0 = 0$.

Differentiating equation (14) wrt t: $\frac{d^2V}{dt^2} = 3\left(H\frac{dV}{dt} + V\frac{dH}{dt}\right)$ ------------ (16).

Now, from equation (14a), $H\frac{dV}{dt} = 3VH^2.$

From the expression for the deceleration parameter we obtain: $dH/dt = -(1+q)H^2$.

Equation (16) then becomes: $d^2V/dt^2 = 3VH^2(2-q)$ ------------- (17).

It would have seemed reasonable to assume that the rate of change of the rate of change of the volume of a coasting universe, i.e. q = 0, would be zero; however, we see from equation (17) that this would only be the case for $q = 2$. The favoured magnitude for q amongst some cosmologists is (or was) $1/2$; we may, for completeness, also consider the situation where $q = -1/2$. The magnitude of the numerical multiplier on the RHS of equation (17) then ranges between 7.5 and 4.5 and has a value of 6 for a coasting universe. Thus, we see, in contradistinction to that which was concluded in **(r2),** it cannot be assumed that a positive value for the rate of change of the rate of change of the volume of the universe is indicative of an accelerating (or, indeed, decelerating) universe.

The cosmic jerk

From the results of the surveys of Schmidt, Riess et al* and Perlmutter et al**, it was reported by Riess in 2003 that the universe went suddenly from a decelerating state to an accelerating state approximately five billion years ago. In Mechanics such an effect is called a jerk.

Whilst the following analysis does not explain the origin of the cosmic jerk it will be shown to provide a way in which the speed of expansion of the universe at the jerk may be determined. The equations developed are somewhat complicated and so, for the purpose of illustration, a rough approximation to the speed at the jerk is calculated.

In view of the above mentioned experimental observations we advance the suggestion that the deceleration parameter q is not a constant thing but is a function of time. In the immediate vicinity of the beginning of the universe it is accepted generally that the universe expanded very rapidly. The acceleration then decreased and the universe entered a phase of deceleration. From equation (9) it may then be inferred that q was very large, and negative, although it is moot whether or not this happened instantaneously. Indeed, it is more realistic to argue that the starting value of q was zero, and hence at inception the universe, instantaneously, was in the condition described by the coasting equations, thereafter accelerating very rapidly. To attain the decelerating state the magnitude of q must have decreased, probably, again rapidly and then passed through zero, in which state, instantaneously, the universe was, for the second time, coasting. Examination of equation (11), which is also equation (14c) with q set equal to zero relates to a coasting universe, and shows that we can set q = 0, at an unknown time before the present, only very slightly less than the inverse of the current magnitude of the Hubble parameter, and still obtain a finite, albeit, exceedingly large magnitude for the value of H at that time. Since it is assumed that the initial phase of increasing and decreasing acceleration occurred very rapidly then this

17

procedure will place the second coasting point in the immediate vicinity of the inception of the universe. Immediately after this time, the algebraic sign of q became positive, corresponding to the universe entering a decelerating phase. It is reasonable to expect that the magnitude of q must have risen to some maximum value and then, according to experimental observations fell progressively, until, approximately some five billion years before the present time, the universe started to accelerate. In order to attain the accelerating state the universe must again have passed through a third coasting condition.

We can attempt to capture the variation in the deceleration parameter q between the magnitude once favoured by some theorists (q = 1/2) and 'now' by assuming that such variation is given by:

$$q = e^{-kt} + B \quad \text{------------------- (18).}$$

We choose this form for the variation of q with t, since previous assumptions of linearity used elsewhere have proved to be less than satisfactory. It is emphasised that the form of the variation chosen is no more than a plausible (but as will be shown later, productive), guess, but at least, with the proper boundary conditions has the merit of changing the sign of the slope of the speed of expansion versus time curve from negative to positive.

Until the observations of Schmidt, Riess and Perlmutter were published the value favoured by some theorists for q was ½. Solely for the purpose of illustration we use this as a boundary condition at a time before the present such that the denominator of equation (14b) becomes arbitrarily close to zero, i.e. at a time very close to $2/3H_0$ before the present time; this state is denoted by the subscript '1'. We must sound a note of caution at this juncture for the reason that equation (14b) is derived on the basis that q is considered not to be a function of time, and so, taking q = ½ at t = $2/3H_0$ has no status vis a vis equation (18), other than that of convenience in the illustrative calculation. A second condition is required in order to determine the magnitudes of, k, and B; this is taken to be the observation reported by Riess that the universe went from a decelerating state to one of acceleration approximately 5 billion years before the present time, and hence, at this point must have passed through the coasting state, q=0.

The condition q = 0 is denoted by the subscript ' 2'. To be consistent with the use of the Hubble parameter expressed in units of s^{-1}, the times in equation (18) must be expressed in seconds, i.e. $t_1 = 0.2773 \; x \; 10^{18} \; s = 2/3H_0$, and $t_2 = 0.1577 \; x \; 10^{18} \; s$.

Now, t_1 and t_2 are in the past of the origin of time, taken to be, now, and hence are negative. Moreover, they also represent differences in time (indeed, there is no absolute time). If we denote 'now' as being the origin of time, in this instance, then we may accommodate this in equation (18) by multiplying e^{-kt} by e^{kt_0}, for this is equal to unity and does not change e^{-kt}. What does change is that the times now become explicitly, time differences, and so, e^{-kt} becomes $e^{k(t_0-t)}$, and the time difference is then positive.

Equation (18) may then be written: $q = e^{k(t_0-t)} + B$ ------------------ (18a).

Using the boundary conditions yields a pair of equations:

$$e^{k(t_0-t_1)} + B = 0.5 \text{ and, } e^{k(t_0-t_2)} + B = 0.$$

Eliminating the constant, B, there results: $\quad e^{k(t_0-t_1)} - e^{k(t_0-t_2)} = 0.5$ --------- (18b).

It should be noted that equation (18b) is completely general (for q = ½ at time, $t = t_1$).

We solve (18b) for $t_0 - t_1 = {}^2/_{3\,H_0}$ by writing it in iterative form, viz.

$$k_{n+1} = {}^{3H_0}/_{2} \, \ln(\, 0.5 + e^{k_n(t_0-t_2)} \,) \text{ ------------- (19).}$$

Preliminary calculations show that this form of iterative equation is robust, and, with a starting value of 10^{-18} the iteration converges after twelve steps to a value of $k = 2.404 \times 10^{-18}$, and hence, $B = -1.4347$.

In this instance, from the above, at the present time ($t = t_0$), we see that the magnitude of the deceleration parameter is, $q = -0.4347$. We then conclude that the negative value indicates that the universe has been accelerating from 5 billion years before the present time. Further, inspection of equation (20) below shows that the nature of the acceleration and its magnitude for given N_{enc} is, in general, dependent upon the group, $({}^{q\,H^2}/_{v})$. Further, each member of this group is a function of time.

It may be noted that we have used a condition, i.e. $t_1 = {}^2/_{3H_0}$ which, if substituted in equation (14b) with q = 1/2 would return infinity as the result. However, if, as previously noted we set t_1 to an unspecified time arbitrarily close to, but less than ${}^2/_{3H_0}$ the infinity is avoided, and this has virtually no effect upon the magnitudes determined for k and B.

We have not attempted to extend the analysis back as far as the second coasting condition after inception of the universe. There is no need to do this for the time range ${}^2/_{3H_0} < t \leq$ 0 encompasses the 'jerk'; in addition, it is doubtful if the simple form of equation (18a) is capable of capturing the variation in q between 'now' and the second coasting point.

We may determine the speed of expansion of space at the jerk, as follows:

The acceleration (or deceleration) of space may be obtained by writing equation (9) as:

$$dv/_{dt} = - {}^{\sqrt[3]{N_{enc}}}/_{2} \, {}^{c}/_{v} \, (\,qH^2\,) = y \text{ ----------- (20).}$$

Using equations (18a) and (14b) it can be shown that the Hubble parameter, H is given by:

$$H = {}^{H_0}/_{\{1 - H_0[(1+B)(t_0 - t) + (e^{k(t_0-t)} - 1)/k]\}} \text{ ----------- (20a).}$$

Inspection of equation (20) reveals that the magnitude of the local frequency of oscillation is required. From equation (7) we may write, $dv/v = -H\,dt$. Integrating yields:

$$ln\,{}^{v}/_{v_0} = \int_{t}^{t_0} H\,dt \text{ --------------- (20b)}.$$

Now, using the substitution $\phi = H_0\left[(1+B)(t_0-t) + \left(e^{k(t_0-t)} - 1\right)/k\right]$, this equation

may be written explicitly, viz. ${}^{v}/_{v_0} = exp\left[\int_{0}^{\phi} d\phi \Big/ (1+q)(1-\phi)\right] \text{ --------------- (20c)}.$

Equations (20a) and (20c) are the underpinnings of a second set of extended equations, similar in form to the previous sets.

Using equation (20) we may write: $v_j = v_0 + \int_{t_0}^{t} y\,dt$ -------- (20c) where, v_j and v_0 are the speeds of expansion of space at the jerk and the current time, respectively. Further, $v_0 = H_0 R_0$.

An alternative method of determining the speed of expansion at the jerk is to calculate the magnitude of H at the jerk from equation (20a), then solving (20c). Since, ${}^{v}/_{v_0} = {}^{l_0}/_{l}$;

then R may be determined and the speed of expansion of space given by the product **HR.**

Without even attempting to manipulate any of the equations associated with either method of the determination of v_j it is considered that an analytical solution is, in all probability not possible. We then must revert to numerical quadrature to obtain accurately the speed of expansion of space at the jerk.

Whilst the process of numerical quadrature is ideally suited to calculation by computer we can, as noted earlier, and solely for the purpose of illustration, make a rough estimate of the speed of expansion of the universe at the jerk, as follows:

The acceleration 'now' is given by: $\left(dv/dt\right)_0 = -\sqrt[3]{N_{enc}}\Big/2 \ {}^{c}/_{v_0}\left(q_0 H_0^2\right) = y_0$

It is shown in **(r2)** that for a universe containing 10^{80} baryons, the number of enclosures (i.e. quantum harmonic oscillators) is, 2.657×10^{92}. The current frequency of oscillation of an oscillator was calculated to be 2.681×10^{12} Hz; $q_0 = -0.4347$, and $H_0 = 2.38 \times 10^{-18}\ s^{-1}$.

Hence, at the present time it is calculated that, $y_0 = 0.8852 \times 10^{-9}\ m\ s^{-2}$.

The acceleration at the jerk is zero but the acceleration of space, according to Riess, has been increasing from the time of inception of the jerk until the present time. If we take the arithmetic mean value, (0) of the acceleration at the jerk and the current acceleration and multiply it by the time elapsed to the jerk, then this will give a rough approximation to the difference between the speed of expansion of space now and that at the jerk. This difference is then subtracted from the current speed of expansion.

We have, for the current accelerating universe, already calculated the speed of expansion of the boundary of space, i.e. $8.55 \; x \; 10^8 \; ms^{-1}$, (2.852 times the speed of light); hence, the approximate speed of expansion of space at the jerk, v_j is:

$$v_j = 8.55 \; x \; 10^8 - \left(\frac{0.8852}{2} x \; 10^{-9} \; x \; 5 \; x \; 10^9 \; x \; 3.154 \; x \; 10^7\right) = (8.55 - 0.698)x \; 10^8 =$$
$$7.852 \; x \; 10^8 \; ms^{-1} \equiv 2.62 \; \boldsymbol{times \; the \; speed \; of \; light}$$

Hence, if the universe had been coasting from five billion years ago to the present time the speed of expansion at the boundary of space now would be that calculated at five billion years ago—this result is confirmed by equation (13a).

There are other implications for this result.

If the universe had been coasting at the present time and the size of the cube of space of this universe remained the same (for the frequency of oscillation of the harmonic oscillator, as calculated from the ground state remains the same), then the Hubble parameter would be smaller than that measured currently by approximately 9 % .

Equations (13) would still be applicable, but the reference Hubble parameter, H_0 would now have the value, $0.916 \; x \; 2.38 \; x \; 10^{-18} = 2.18 \; x \; 10^{-18} \; s^{-1}$. It should be noted however that this magnitude is within the tolerance band of the currently-accepted value of Hubble's parameter.

Now, equation (18a) is only valid within its temporal range and extrapolation into the future is not attempted.

There is a further and important consideration that must be addressed.

The placing of $q = \frac{1}{2}$ at a time $\frac{2}{3} \, H_0$ b p arose from equation (14b) which was derived by taking q to be a parameter. Now, the magnitudes of quantities derived from this may only be gross approximations, in the sense that it seems more reasonable, in view of q being some function of time, to assume that the temporal position of the deceleration parameter $q = \frac{1}{2}$ is at some point other than $\frac{2}{3H_0}$. We have the opportunity to test this contention using equation (20a).

The magnitudes of k and B are dependent upon what time is chosen for t_1 in the time difference, $t_0 - t_1$. If the term $H_0 \, [- - - - - -]$ in the denominator of (20a) becomes unity then the magnitude of H becomes infinite. In essence this is same approach that was employed in examining the denominators of equations (11) and (14b to determine the time difference, $t_0 - t$, at which H became infinite; although in this instance the procedure will be somewhat more complicated. We now proceed to determine magnitudes for k and B which will allow the movement of the theorists favoured point $q = \frac{1}{2}$ closer to the temporal point consistent with the variation expressed in equation (18b).

We write $t_0 - t_1 = {}^x/_{H_0}$, where $0 \leq x \leq 1$ and rewrite equation (19) as:

$$k_{n+1} = {}^{H_0}/_x \ln(0.5 + e^{k_n(t_0 - t_2)}) \text{ -------------- (21).}$$

If we repeat the iteration for **x = 1, 0.98, 0.95, 0.9**, it is found that the denominator of equation (20a) has the values; **-0.0808, -0.0723, -0.0376, 0.02847**. It then follows that the magnitude of the denominator has passed through zero, and by further calculation we can determine, to good approximation, the value of x at which this occurs.

When **x= 0.925** the magnitude of the denominator is **-0.004645**. We then conclude that the x associated with the 'zero point' is very close to this value. We could repeat the calculation, progressively refining the value of x until the magnitude of the denominator became vanishingly small but this is regarded as a somewhat profitless exercise. Accordingly, we accept that the value of the deceleration parameter was ½ when the time b p was $0.925/_{H_0}$ (i.e. 12.3 billion years b p). The corresponding values of k, B and q_0 are $1.45 \ x \ 10^{-18} \ s^{-1}$, **-1.257 and -0.257**, respectively. Using these results, and repeating the approximate calculation shown before it is found that the speed of expansion of space at the jerk is calculated to be **2.79** times the speed of light.

Now, we have described previously a possible distribution of the deceleration parameter in the vicinity of the inception of the universe. The distribution consisted of a coasting point at inception then a very rapid acceleration followed by an equally rapid (more or less) reduction in acceleration which passed through a second coasting point and entered a deceleration phase, rising to a maximum value and then proceeding to decrease. If we take the age of the universe to be given closely by the Hubble time then we see from the above calculation that

q = ½ is positioned a significant distance in time away from the start of the universe. The temporal position of q = ½ is determined by the calculated value of x which will result in the denominator of equation (20a) becoming zero. But, this only expected to occur at the inception of the universe. Hence, it is a distinct possibility that q is not equal to ½ and that it may be positioned arbitrarily closely to the Hubble time. This implies that the first and second coasting points are in close proximity to each other.

All of this may be investigated by rewriting equation (21) as:

$$k_{n+1} = {}^{H_0}/_x \ln(q + e^{k_n(t_0 - t_2)}) \text{ -------------- (22).}$$

We may then use the same approach as that employed to establish the position of q at $0.925/_{H_0}$, but using values of q other than 0.5. Given the tedium of the calculations involved this will not be undertaken here.

Equation (20a) is considered to be a more realistic expression for the variation of the Hubble parameter with time, albeit only valid within the range of the variation for q.

However, for the purpose of illustration, the variation of H with time for a coasting universe will be employed in determining the variation of the temperature of blackbody radiation with time.

Now, it can be shown that the current temperature, θ_0 of blackbody radiation is related to the temperature at an earlier time by: $\theta_0 = a\theta$, (strictly, only applicable to photons), where a is the scale factor, and which, from equations (13) is given by $a = {}^{v_0}/_v$, setting a_0 equal to unity.

From equations (13) we have, ${}^v/_{v_0} = {}^H/_{H_0} = {}^1/_{[1 - H_0(t_0 - t)]}$

Hence, ${}^\theta/_{\theta_0} = {}^1/_a = {}^v/_{v_0} = {}^1/_{[1 - H_0(t_0 - t)]}.$

It then follows that equations (13) may be expanded thus:

$${}^v/_{v_0} = {}^H/_{H_0} = {}^{a_0}/_a = {}^{l_0}/_l = {}^\theta/_{\theta_0} = {}^1/_{[1 - H_0(t_0 - t)]} \text{------------ (23).}$$

The temperature of blackbody radiation, θ_j , for example, at the time of the cosmic jerk, as reported by Riess, is :

$$\theta_j = {}^{2.725}/_{[1 - 2.38 \, x \, 10^{-18} \, x \, 5 \, x \, 10^9 \, x \, 3.154 \, x \, 10^7]} = 4.362 \, K.$$

Thus we see that, even for the very simple model of a coasting universe, the temperature of blackbody radiation five billion years before the present time was greater than, but not substantially different from, the current temperature.

Now, $a = {}^1/_{1 + z}$, where z is the cosmological redshift; hence

$$1 + z = {}^1/_{[1 - H_0(t_0 - t)]}$$

The variation of the redshift with time for a coasting universe is then given by the formula:

$$z = {}^{H_0(t_0 - t)}/_{[1 - H_0(t_0 - t)]} \text{---------- (24).}$$

Further, it may be shown from, $a = {}^l/_{l_0}$ [equations (23)], and $a = {}^1/_{1 + z}$, where z is given by $z = {}^{(l_m - l_r)}/_{l_r}$, the subscripts m and r denoting 'measured' and 'reference', respectively, that:

$${}^l/_{l_0} = {}^{l_r}/_{l_m} \text{------------ (25).}$$

All of the foregoing may be extended by incorporating the complexities of the previous section.

The Hubble parameter.

In the following we show that the Hubble parameter may be assigned meaning other than a proportionality parameter relating the speed of expansion of space at a point to the distance between that point and an observer.

A slight rearrangement of equation (14a) shows that: $\frac{1}{3V}\,\frac{dV}{dt} = H$.

This result is not dependent upon the dynamic of the universe and hence has general application.

In words: the Hubble parameter may be interpreted as, one third of the temporally-local rate of volumetric strain. This result is applicable to a fundamental cube as well as the entire universe.

If we form the product, $p\,\frac{dV}{dt}$, where p is the vacuum pressure, then this may be interpreted as the rate at which work is being transferred in the expansion of the space of the whole universe, assuming, of course that, in this instance the volume in question is that of the whole universe.

As noted in **(r1)**, following Davis and Griffen, **(r4)**, it is assumed that the vacuum, taken to be devoid of any dissipative processes, behaves as a perfect fluid, and, to maintain Lorentz invariance should also have no preferred direction. Hence the first term in the energy tensor for a perfect fluid must be zero, requiring that the vacuum pressure be equal to the <u>negative</u> of the energy density of the vacuum, taken in our model to be the ground state energy density, ρ_E, of the quantum harmonic oscillator.

We may then write: $p\,\frac{dV}{dt} = -\rho_E 3\,V\,H$.

In all of the previous work, the ground state energy density, is given by: $\rho_E = \frac{h\,v}{2l^3}$.

Now, the volume of an ensemble of fundamental cubes is given by: $V = N_{enc}\,l^3$, hence we may write:

$$p\,\frac{dV}{dt} = -\frac{3}{2}\,h\,v\,N_{enc}H \text{----------------- (26).}$$

It is shown in **(r1)** that the total negative energy (incorrectly called the Dark energy), D, of the universe at any given frequency is: $D = -\frac{3}{2}\,h\,v\,\hat{N}_{enc}$, where $\hat{N}_{enc}$ is the number of enclosures in the universe, and so we may write: $\frac{1}{D}\,p\,\frac{dV}{dt} = H$.

In words; irrespective of the number of fundamental cubes in an ensemble, from equation (18) it follows that the temporal rate at which work is being transferred to space from the reservoir of negative energy, relative to the total quantity of negative energy, is equal to the local magnitude of the Hubble parameter. The inverse of this ratio is the local magnitude of the Hubble time.

It is important to note that this definition only applies in the case where the vacuum is capable of falling into the ground state, and hence strictly obtains only during a free vacuon mode. In addition, since H is positive then the rate at which work is being transferred to space is negative, and so, space is expanding.

Further, it is of interest to note that in (r2) it was shown that, in a free vacuon phase a meaning could be assigned to the ground state energy density such that it could be regarded as the ratio of the rate at which energy was being added to an enclosure relative to the rate at which space was expanding.

In view of all of the foregoing, it is has been shown that it is more appropriate to regard the Hubble parameter as a function.

The expansion of space relative to an observer.

In this section aspects of the act of observation are considered.

Before proceeding, we establish an important caveat which often seems to be overlooked when considering the observation of an event.

It is posited that a point observer may only look in one direction at any given time. For such an observer, the concept of a 'field of view' has no meaning, because images of multiple, spatially-separated events impinging simultaneously upon a single recording point would render discrimination between the events impossible. In order to observe an event in another direction, the point observer must look in the direction of that event. Further, to observe multiple events in the same direction the events must be temporally- separated.

The purpose behind the foregoing is to establish the primacy of the point at the centre of a cube, and thus the location of a point observer. Indeed, by locating an observer within the space being created removes the illogicality of viewing the creation of space externally.

Consider the plan view of a cube, such as analysed in the previous section, with an inscribed sphere, as shown in Fig4 on the diagram sheet, and, let a point observer, O, (as described above) be located at the centre. If, for example, the observer looks to the North, he/she will, due to the expansion of space, see a test object currently occupying the position at distance, R, to be moving away at speed, v, relative to the observer. In order to see in the directions, East, South and West, the observer must look in these directions, and again, will see relevant test objects at distance, R, to be moving away at the same speed, v.

Now, the orientation of the sphere/cube combination is arbitrary, and so, if we now require the observer to look out to the same distance as before in the directions NE, SE, SW, NW, we may ensure that, again, the observer sees that the points observed are moving away at speed, v, by rotating the cube, as shown in Fig5. This procedure is valid in any direction in three-dimensional space, so that an observer looking out from the centre of a given cube in the direction of a normal to a side of the cube, will always see a test object embedded in the plane of space coincident with that side of the cube to be moving away at speed, v. As shown previously, it follows that the observer will see space expanding in all directions at a speed related to the distance from the observer to the observed point, irrespective of the position of the observer in space.

The observation made by an observer that all points in space are moving away from him/her implies that the observer is moving with a point in expanding space and hence the point/observer combination constitutes a co-moving coordinate system.

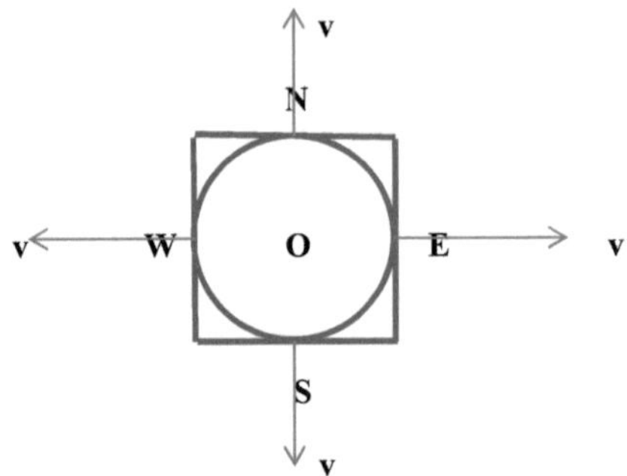

Fig 4

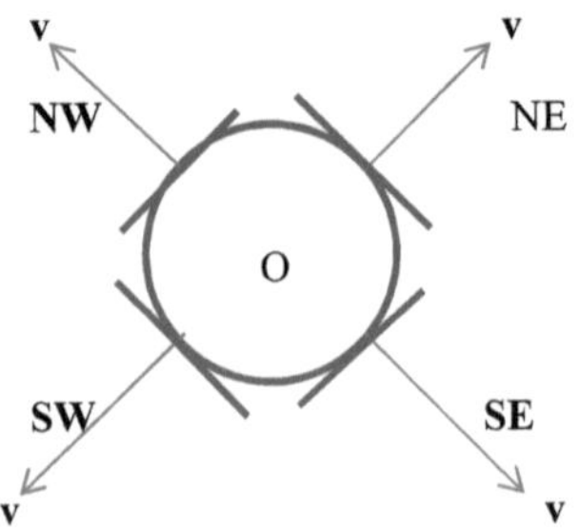

Fig 5

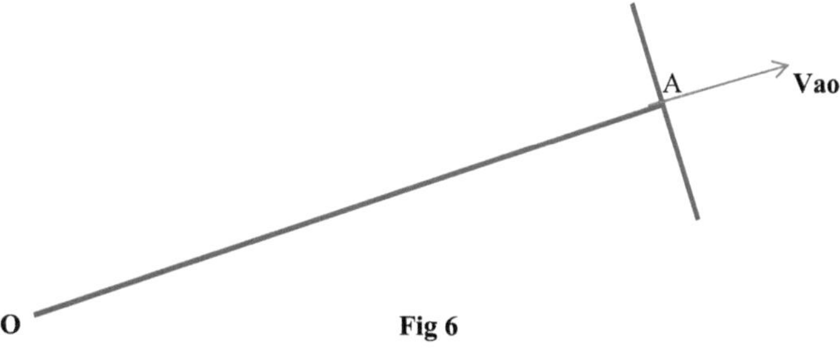

As shown in Fig6, we may dispense with the idea of a complete cube and sphere and replace the depiction of the geometry of an observation by specifying the location of the observer with respect to some event, drawing an arc, the radius of which is equal to the distance of the event from the observer and, at the point of intersection drawing the tangent (which is, in fact a tangent plane) to the arc. Whilst it is obvious that the direction of the expansion of space is normal to the tangent, a diagram of this nature also shows clearly that if the observer moves to the location of the event and measures the speed of expansion at her original location then she will observe space to be expanding away from her at the same speed as before. Of course, and as mentioned before, the expansion of space is not directly observable and what is meant is that, in this context, an event consists of the measurement of the wavelength of radiation emitted from a material object being carried with the Hubble flow at some fixed distance from the observer. If space is expanding as described, then, in all directions of observation the radiation will be seen to be red-shifted. If, due to some unspecified effect a point is seen to be moving towards the observer, then the effect is locally overcoming the effect of the Hubble flow and the emitted radiation will be seen to be blue-shifted.

From all of the foregoing, we recognise that, in the context of the expansion of space, a stronger version of the Copernican principle is that no point in space is special and, concomitantly, neither is any direction of observation.

It is then concluded, for this model, that the universe is infinite and is observer-centred; each observer is at the centre of his/her cube of space and can only observe that part of the universe. Unless the positions of two observers coincide, then each observer, could, in principle, see a different part of the universe. However, if we take, for example, a group of observers located at the current time on the surface of a sphere, 8000 miles in diameter, it is easy to show that all of these observers do not possess sufficient resolution to discern any difference in their observations when observing their parts of the universe on cosmological scales, for, if we take two observers located at the opposite poles of the sphere, the orientation of the line joining the poles being completely arbitrary, then, for the size calculated for the universe at the current time (a cube of side 76 billion light years) the ratio of the distance of separation for the centres of the cubes of each observer relative to the side dimension of a cube is calculated to be of the order of 10^{-20}. Hence on cosmological scales the centres of each cube are virtually coincident, and it is highly unlikely that each observer will possess the precision necessary to distinguish any differences in their far-field observations, if indeed, such differences exist.

Finally, it is contended that the Hubble flow throughout the universe is the same as that of the part examined here.

Discussion

Rearranging equation (25) gives: $l\,l_m = l_r\,l_0$. This expression has an interesting meaning, for, what it states is that the product of the wavelength, l of the vacuum at some time in the past and the wavelength of the radiation, l_m emitted from a source carried with the Hubble flow but measured now, is equal to the product of the wavelength, l_0 of the vacuum now and the wavelength, l_r of the emitting substance measured at the present time. In addition, the left hand side of equation (25) relates entirely to the vacuum whilst the right hand side relates entirely to the emitting substance. Indeed, equation (25) can be argued to have general application for it is not based upon any particular variation for the deceleration parameter, q.

The equation has an even more interesting interpretation than that described above, for it states that, the ratio of the wavelength of the vacuum 'then' relative to that of the vacuum' now', is equal to the ratio of the wavelength of emitted radiation 'then' to that of the radiation measured 'now'; in this we have made the perfectly reasonable assumption that the wavelength of the emitted radiation 'then' is the same as that at the reference condition, 'now'.

W M Fidler   February 2016.

References.

(r1) The Dark energy co-existing with the vacuum of a universe, modelled as a set of identical, frequency-quantised, simple quantum harmonic oscillators.

W M Fidler September 2013 (unpublished)

(r2) On the evolution of the mass/energy of a universe from the vacuum, modelled as a set of identical, frequency-quantised, simple quantum harmonic oscillators.

Addendum: On the nature of Dark Matter---a neutronic ring theory.

W M Fidler July 2014 (unpublished)

(r3) The determination, for a universe where the vacuum is modelled as a set of identical, frequency-quantised, bi-modal, simple quantum harmonic oscillators, of:

1. The temporal variation of the mass density during the second sterile phase.
2. An estimate of the time interval for the formation of the Dark and baryonic matter, together with a rough approximation for the time interval for the formation of the quark/gluon plasma.
3. The temporal variation of the mass density during the second free vacuon phase.
4. The rate of creation of space and the acceleration of this rate during the second free vacuon phase; conditions at the cosmic jerk.
5. The volumetric expansion rates during the second free vacuon phase.
6. The magnitude of the current photon/baryon ratio.
7. The different versions of Einstein's Field Equations.

W M Fidler April 2015 (unpublished)

(r4) The Cosmological Constant

Tamara Davis & Brendan Griffen.

Scholarpaedia.org/article/cosmological- constant. October 2011.

*Observational Evidence from Supernovae for an Accelerating Universe and a Cosmological Constant.

The Astronomical Journal **116(3):** 1009-1038.

**Measurements of Omega and Lambda from 42 High-Redshift Supernovae.

The Astronomical Journal **517(2):** 565-586.